MW01469085

Quantum Physics for Beginners

The Simple and Easy Guide

in Plain and Simple English

Without Math

Donald B. Grey

Bluesource And Friends

This book is brought to you by Bluesource And Friends, a happy book publishing company.

Our motto is **"Happiness Within Pages"**

We promise to deliver amazing value to readers with our books.

We also appreciate honest book reviews from our readers.

Connect with us on our Facebook page www.facebook.com/bluesourceandfriends and stay tuned to our latest book promotions and free giveaways.

Contents

Introduction

Life, time, and the universe are normally understood to be linear. Everything happens in a certain sequence and can't be altered once certain actions have been made, and you can only be in one place at one time. This, although taught for the majority of the scientific fields, does not agree with the quantum side of this scale.

Quantum physics was first introduced in the 1920s by a group of scientists like Niels Bohr, Erwin Schrödinger, Werner Heisenberg, and Albert Einstein. There were, of course, many other scientists that contributed to this field of science, but these four scientists are some of the more famous ones.

Quantum physics may be a difficult area of science to understand, but, put simply, it is the branch of physics that specifically focuses on particles, how particles interact, and how every conceivable thing in this universe (or multiverse—but we'll get to that later) is made up of these astronomically small things. Quantum physics focuses on atoms and how they interact with each other; because they focus on the building blocks of all matter, quantum physics also has a direct influence on other fields of science like life sciences (biology) and chemistry.

Everything that you will need to explain revolving around a scientific field will need some backing from quantum physics. Even basic principles of how electricity and electrons move through electronic chips, how photons in light can be magnified and amplified to create lasers, and even the fact that a star can keep burning for billions of years, all have a direct link to quantum physics.

Even though every known area of science has some form of quantum backing, it is still one of the most difficult fields in science to wrap your mind around. It may be difficult, but if you're reading this book, you also understand that there is definitely a certain amount of attractiveness and curious fun about this branch of science. Diving into this field will expand your mind in places that you didn't even know were possible.

Maybe you have watched Christopher Nolan *(Memento)* movies and walked out of the cinema with your mind reeling, trying to process all of the details that you just saw. If you are attracted to quantum physics, then movies like *The Prestige, Interstellar,* and *Inception* or series productions like *Fringe,* or *Flash Forward* will have truly tickled your fancy. The beauty of these movies is that they all bring about certain aspects of quantum mechanics and how it influences other areas of the world and human beings themselves.

I can personally say that, after I watched *Interstellar,* I sat back on my couch and pondered about the whole movie for about thirty minutes. I then proceeded to watch the movie again… Even though it is a considerably long movie. Sometimes, mind-blowing ideas need to be mulled over several times before the concepts start to sink in. This process of understanding your inner curiosity is merely the beginning steps to understanding the stages of quantum physics.

What makes quantum physics so confusing is that it opens up the idea of potentially different dimensions to our universe. This means that there could be parallel universes that are running concurrently with ours. Every decision that we make in our lives will result in repercussions, some big, some small. But each action will have a specific reaction and will influence the dimension that we are all experiencing at this time. This is where the butterfly effect and the chaos theory are birthed (both of these will be detailed on and discussed later).

All the scientists that have had a major influence in this field may agree with each other's theories or disagree completely. The problem here is that, because quantum physics is such an "out there" field, some pretty insane theories can be developed. That said, it does not mean that these theories are incorrect.

Einstein was one of the scientists who understood the possibility of time warping, wormholes, and even probable travel into the future. He realized during his research that light would have a direct impact on his theory. He realized that because light travels at over 670 million miles per hour, speed could have a direct bearing on the effects of time. While developing his theory of relativity, he concluded that traveling at the speed of light could, in a way, warp time.

An example that he made was that if a train track could be built around the world and a train could be developed to travel close to the speed of light (I say close to because nothing can ever go as fast as light itself), and travel without cessation... At that speed, time inside of the vessel would, in a way, slow down, while time outside the vehicle would remain at its constant linear speed. So, if people were inside that train as it approached the speed of light, they would age at a much slower speed than the people outside of the train, because their time would slow down considerably. After a month of traveling at that speed, the people may climb off the train and see that the world could have aged years during the same interim.

Einstein's theory of relativity (which will be discussed in depth later), may be very difficult for many people to understand, and it did cause some waves in many scientific circles. Nevertheless, it created much further impact in quantum physics and other branches of science.

This scientific field may be overwhelming to some, but you have been curious enough to look into it. This may change your whole outlook on a much grander scale.

This book does not have tons and tons of mathematical formulas for you to go through, as a typical physics book does, which can make it very difficult to read. Even with the "impossible math," after reading this book, you will understand the basic concepts of quantum physics, and your mind will be opened to a whole new world you never thought possible before.

Chapter 1: What is Quantum Physics?

Let's face it… Quantum physics has to be one of the most confusing topics on the face of the planet, but it is also one of the most intriguing, hence why you are reading this book. Quantum physics seems rather strange and, in a way, counterintuitive because of how different it is to normal principles that are taught in most other scientific fields. And, truth be told, most of the scientists that deal with this field in physics find it rather confusing and tiresome most of the time too. The reason for this is simple—many of the measurements in theories in this field seem almost unquantifiable because of how "out there" many of these theories are.

That said, the measurements and theories in quantum physics are not completely incomprehensible, and many of them have opened the door to further understanding within other aspects of science. Fortunately, to understand how quantum physics came about, and why it plays such an integral part in the world that we live in, can be broken down into a few key elements that will help you wrap your mind around this difficult concept.

Everything in This World and

Beyond Is Made up of Particles and Waves

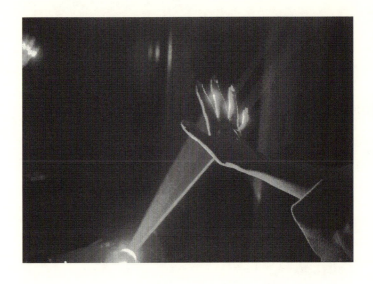

Before we continue, you need to understand that everything in this world and in the universe that we know, is made up of both waves and particles. This may seem like a strange concept because it is generally understood that objects cannot be both. Light generally moves in waves, while tangible matter is made up of particles. According to traditional science… These two concepts don't exist together in the same object.

Quantum physics, on the other hand, says that everything in the universe that we know of has properties of both waves and particles. This goes against traditional physics. In fact, most of the concepts in the quantum realm do go against traditional science. This may seem very confusing and may seem like a rather insane notion, because describing real-life objects as a combination of both particles and waves can be rather inaccurate.

In more accurate terms according to quantum physics, the different types of objects and their makeup are actually neither waves nor particles, but seem to move into a different category altogether. This different category moves objects into a place where they share properties of both waves and particles, and are influenced by both.

The reason why both of these aspects need to be included into objects is because waves can generally be spread over a certain distance or degree, while particles can be compressed and localized. Many objects in space and time may share different aspects of these types of categories, but may not fit into one category entirely.

Although most objects do share qualities of both, there are still many circles in scientific fields that refuse to accept this idea. One of the most commonly seen debates is when certain teachers and physicists claim that light is made out of particles... Which creates certain contention with some of their peers. The reason why this is a rather debatable topic is because it can cause some confusion between people who are still learning basic concepts in physics. Of course, light does have a nature that acts towards particles instead of waves, because light is made up of accelerated photons. Photons are, in a way, particles, even though they are constantly moving.

This can, unfortunately, lead to some misconceptions: Even though photons are particles in nature, they act completely differently to what particles are supposed to. In the quantum field, they are rather referred to as "excitations", instead of "particles" or "waves", because they are, in fact, particles with wave-like properties.

Quantum physics allows for a much greater and deeper meaning of the universe because it combines different properties from different states at a single time. While it greatly adds to the complexity to the universe, it also allows for a grander form of intricate beauty.

Quantum Physics Discreetly Hides in the Background

If a person dives into the meaning of the word "quantum", they will quickly find that it means "how many or how much," when directly translated. This is very appropriate because of all of the quantum models that have been designed from the 1920s to present. It shows that there is a certain amount of a specific substance or energy-form that will come in discrete or subtle amounts. Even though these amounts are considered discreet, they are still multiples of larger integers and exponents (in some cases) to create fundamental forms of energy.

This may sound confusing, but it all boils down to the different types of wavelengths that a certain frequency of light has. High frequencies have shorter wavelengths but are characterized by their high amounts of energy. Low frequencies have longer wavelengths and are characterized by their lower amounts of energy.

Unlike other forms of wavelengths and energy concentrations, the total amount of energy that is contained in these light fields is multiplied according to an integer amount (integers are whole numbers and do not include fractions and decimals). This is one of the primary properties of numbers within a quantum field.

One of the areas that shows how quantum physics works in a discreet fashion is by the implementation of atomic clocks. Atomic clocks are accurate down to a millisecond and are precisely accurate. Atomic clocks are ultimately this accurate because of the quiet and discreet backing of quantum physics.

Using the understanding of quantum physics and how the frequency of light is measured according to integers, the transition between the light frequencies in the two allowed states in the cesium in the atomic clocks can be almost absolutely accurate. Because of the understanding of the light frequencies in atomic clocks, these clocks can be so accurate that they barely deviate by a second every 100 million years.

Quantum Physics Works According to Probability

One of the aspects that is constantly debated by learned professionals and basic dabblers in quantum physics alike, is that many believe that outcomes in quantum physics experiments are almost impossible to predict—especially if they are performed within a quantum system instead of a closed system. The difference between the two is that a closed system does not allow for external influences during an experiment, while a quantum system occurs when two or more quantum states interact with each other (this will obviously affect outcomes dramatically, and may make it unlikely for the experiment to be replicated).

When scientists make a hypothesis regarding a certain experiment, they predict the outcomes of the said experiment through the use of mathematical probability. Once the experiment has been completed, comparisons will then be made between the beginning hypothesis and the outcomes of the experiment, and they will then infer the probability of their predicted outcomes (obviously adjusting accordingly after the experiment is completed and then repeated).

Since quantum physics has a lot to do with light frequencies and wavelengths, it is only understandable for experimental quantum systems to be based on these same principles. These experiments are based on the wave functions of the different types of light exposures and how it will affect the probable outcome of the experiment.

If you are sitting on your couch at this moment and wondering what wave functions actually are... Don't worry, because most learned professionals aren't entirely sure about it either, and keep debating on it. However, what can be deduced from their arguments is that wave functions are either physical entities that can be measured and predicted, or that wave functions are merely an expression of the amount of knowledge that a certain person has about the outcomes of a certain experiment based in a quantum system.

Scientists use this understanding of wave function (or the idea thereof), to try and predict the outcomes of a certain experiment. After years and years of debating what wave functions actually are, scientists are still struggling to agree on this front. They have generally agreed on the notion that wave functions can allow particles to be in multiple states at the same time (opening up to the possibility of alternate dimensions).

However, this still remains hypothetical because these types of predictions are practically impossible (or improbable, depending on whom you ask). All scientists can do within systems like this is to predict the potential probability of a certain experiment turning out in a certain way.

The experiment will then be judged according to the scientists' preconceived idea or notion regarding wave functions, and it will try to measure a particular outcome. Basically, a scientist will be able to judge if his hypothesis was correct after observing a particular outcome from his/her experiment.

The quantum system that is ultimately being tested and measured will end up in an undefined state. This system can both mathematically and scientifically map out all of the particle's potential positions in different dimensions (an object's superposition). Once its superposition has been taken into account, potential probabilities of the outcomes can be calculated.

Scientists normally end up at an impasse here. Some believe that the experiment in the system is actually experiencing all states at a single moment, while others believe that the experiment is merely part of one larger "unknown" state. You will quickly come to see that, although quantum physics may be the basis for many other scientific fields, it is also one of the least understood and largely debated scientific fields in the world. The main reason is that many of these tests do not actually make sense... Until they do. For many scientists, experiments in quantum systems often feel like shots in the dark.

Quantum Physics Cannot be Localized and Does Not Have a Specific Location

Albert Einstein had one of the world's greatest minds in existence. He designed theories and solved problems that many scientists could only dream of (some of which will be discussed later). Nevertheless, he was human and also made mistakes with some theories.

That said, Einstein's final contribution to his field may have been incorrect. There was a problem revolving around the quantum side of physics that was plaguing his mind, and it was not something he could leave alone until he dove into it. So he and two younger colleagues decided to write a paper revolving around "quantum entanglement."

Albert Einstein, Nathan Rosen, and Boris Podolsky joined efforts to write the EPR paper (the abbreviation deriving from the initials of their surnames). This paper argued that quantum physics allowed the existence of other dimensions and systems. These systems were measured at separate locations and were quantumly-connected to each other. This means that one directly influenced another, and one's primary outcome was determined by another system.

They argued that all measured outcomes need to be measured ahead of time. They realized that they had to be able to predict the outcomes of certain experiments that had bearings on other outcomes. If they had to wait for the data to be transmitted from another location, they would need a way to be able to transmit that data at the speed of light for accurate results (which was not humanly possible at that point in time).

Because of the gaps in accuracy from these different types of tests, it was derived that there was a large gap in quantum mechanics. They believed that the quantum field was merely an approximation of a much larger and deeper theory.

Since there was no one to actually contradict their statements, for decades to come, the EPR paper was used as a footnote in many quantum papers. That was until John Bell, an Irish scientist, dove into the EPR paper and decided to test its details and claims.

After intensive testing and research, John Bell concluded that it is possible to find specific circumstances where quantum physics can accurately predict correlations between measurements that are separated over a distance. These correlations proved to be much stronger than any of the theories that were stated in the EPR paper. These correlations were eventually tested in the 1970s by John Clauser, and in the 1980s by Alain Aspect. It was conclusively proved that the "entangled" systems, as described in the EPR paper, could not be explained by any hidden variable theory that was localized.

It is now understood that quantum physics cannot be localized. The results of measurements that are made at certain locations will specifically depend on the properties of other influential objects that are a distance away. These objects use signals moving at the speed of light.

That said, this does still not allow information to be sent at the speeds that exceed the speed of light. This has been unsuccessful, even though there have been attempts to use the properties of quantum non-locality to accomplish it. The discussion of time-warping and wormholes will be discussed in the chapters to come. Theoretically, quantum non-locality can achieve speeds that could potentially move faster than previously imagined possible.

Quantum non-locality is very important in this area of physics because it increases the probability of black holes being able to lead into other dimensions. To make things even more interesting, it's believed that wormholes can warp space and time. This idea produced some very radical mathematical ideas that claim wormholes have a structural mathematical identity to them and are theoretically plausible.

This opens up a whole new field in quantum gravity studies about potentially gaining the ability to time travel or teleport—but we will get to that later.

Quantum Physics is Actually Smaller than You Realize

Don't misunderstand this statement. Quantum physics is an integral part of all branches of science... But it is based on very small particles. Quantum physics is weird because the assumptions and predictions in it are very different from everyday experiences.

It is ultimately weird because within a quantum system, a person wants to see that particles can react like waves or vice versa. All phenomena that occur within the quantum realm are generally confined to a very small scale—an atomic scale, to be precise. Because of these atomic effects, the wavelengths and energies that are released from the behavior of these atoms and particles will have visible results in the real world. As you will notice with theories like the Chaos Theory and the Butterfly Effect, small changes in behavior will have much larger influences and effects later on in time.

These changes allow for exponential growth, similar to how certain toxins can multiply when an organism feeding on them gets bigger. For example, mercury concentrations increase dramatically in larger fish in the oceans of the world. Smaller fish still have low levels of mercury, but if that small fish is consumed by a larger fish, the mercury concentrations increase in the larger fish. If that fish is then consumed, the mercury concentrations will increase again. This exponential growth rate is one of the reasons why people shouldn't eat large fish very often.

Quantum physics works in exactly the same way. Minute changes at the beginning of an event can lead to exponentially large outcomes at the end. Quantum physics may be small but that doesn't mean that it is any less influential.

Quantum Physics is Science not Magic

Quantum physics may be very weird, but it is not magical. The predictions that it makes are considered strange when compared to traditional physics. But, they still abide closely to the rules and standards set in place by both mathematics and scientific laws. So, if anyone conjures up an idea that is claimed to be on the "quantum" side of things, and it seems too good to be true… Then it most likely is.

That said, it does not mean that quantum physics can't be used to predict and allow for mind-boggling and amazing things. But, it does need to remain within the laws of science and common sense. Just because time travel is potentially theoretically possible, that doesn't mean it is achievable—well, not yet anyway. We haven't yet managed to break that unknown barrier.

There are aspects about this scientific field that are incredibly difficult to understand, much less calculate. But it does still remain within the laws of science, even if those laws are not completely understood yet.

It is necessary to remember that quantum physics is still science, and is still capable of blowing your mind while revolutionizing some aspect of the world. But there are things that are still highly improbable to occur, even if not completely impossible.

Chapter 2: The Chaos Theory

In 1974, in the early hours of the morning in the small town of Los Alamos, New Mexico, the police were called with regard to a very suspicious man who seemed to be prowling the streets. They eventually figured out that this man was Michael Fagenbaum.

He was trying to find a way to increase the hours in his day to 26, instead of the usual 24. For obvious reasons, this caused several of his peers to question his sanity, and as it turns out, they had on multiple occasions.

Michael Fagenbaum was definitely a strange type of physicist and was not well-known in any sense of the word. He had only published one article, and he always seemed to be attracted to projects that didn't look very promising. At the age of 29, he looked more like an older, eccentric German composer than a young man his age. He struggled to get enough work done in a 24-hour period, so he tried to find ways to warp time. He would walk the streets at night (hence the police encounter) to try and solve the problems that were plaguing him.

All of Fagenbaum's colleagues were very curious about his aloofness and unwillingness to try and solve problems and produce work of his own. He kept up his curiosity, even though he didn't voice many of the problems that plagued him. He watched how liquids would erratically respond when he poured milk into his coffee, and noticed how clouds kept random sequences and still kept a form of non-randomness—almost like the sulci (the grooves) and gyri (the ridges and folds) that make up the hills and the valleys in the brain.

He started to notice that there seemed to be a mathematical construct behind all the randomness. What Fagenbaum's colleagues didn't realize was that he was working on something quietly. And it was something that ran much deeper than any of them imagined. He was working on Chaos.

Where chaos takes form, classical physics stops. Physicists (before 1974) dedicated their time to study the laws of nature, but had not noticed how random the weather patterns were. There were a few scientists in the 1970s who broke out of that ignorance and realized that there was much more to be understood about aspects of the physical and non-physical world.

Chaos became a science in itself, and back in the 1970s, when computers weren't nearly as advanced as they are now, this new theory was changing many aspects of understanding the world. The physicists that were open to this new type of thinking realized that there was a lot more to physics than just the classical natural laws. Many physicists claimed that it was like someone had found the light-switch to the universe and turned it on.

Chaos was a new science that had created a new language for itself. These new constructs were known as fractals, smooth noodle maps, bifurcations, folded-towel diffeomorphisms, intermittencies and periodicities, and all became part of the language of the construction of chaos and the understanding of its true nature. Don't worry, you won't have to learn about all of these aspects to understand the basics of chaos!

Many physicists soon realized that chaos was more of a journey than a destination. It is the science of progressively processing instead of being in a single state. This meant that they could see chaos in the behavior of the world's weather patterns, traffic, or the liquid being pumped through pipes underground. Once the physicists realized that chaos was all around them, they knew that they needed to find ways to mathematically predict outcomes for some of the random structures that developed around them.

No matter what the device or medium, chaotic behavior will follow the newly discovered laws and principles within the random constructs of this theory. Chaos caused divisions in scientific communities and obliterated the lines that separated most of the scientific fields and disciplines. Before the development of this theory, scientific fields were carrying on without much development. But the introduction of this theory changed everything.

Chaos opened up a whole new can of worms and presented new problems that defied the previously accepted ways and laws. It showed that there was a lot more to the universe and how it worked than meets the eye.

Physicists became divided very quickly after the introduction of this theory. Some believed that understanding chaos was so important that it rendered other fields in science completely obsolete. Others went so far to say that the only three fields in science that would be remembered from the 20th century would be quantum physics, relativity, and chaos.

There are aspects of physics that seem to work themselves into a corner, and over the last 40 years, scientists have gone back to the understanding of chaos to find a way out of that corner. Physics has shown us the laws that govern our lives, the world and our universe, but there are still areas in life that physics cannot predict.

However, being able to predict certain events is what gives quantum physics and chaos their importance. Being able to predict certain outcomes changed the world and produced many different types of advancements in the field. Physicists realized that, in a universe ruled by entropy and chaos, things would eventually lead to greater forms of disorder. It's important to understand how organized constructions can still arise and thrive, but because of chaos, even simple systems can create very difficult problems, which can make predictability highly unlikely.

That said, predictable order still arises in those systems and seems to function without problems. It is like chaos and order are good bedmates that can work together for the sake of random predictability. Scientists observed these chaotic and orderly constructs within the same systems and realized two key factors: First, small differences in the input in the beginning stages of an experiment can (and will) result in greater and more noticeable effects towards the end of the experiment. Second, chaos is merely the beginning of chaotic theories. Another one was quickly developing as a result: The Butterfly Effect.

The Butterfly Effect

It's a strange notion to say that a butterfly flapping its wings could accelerate a sequence of events that could ultimately lead to a hurricane on the other side of the world within a few months' time. It may sound impossible, but once a person starts to understand that small actions at one point in time can result in much larger consequences later on, then the butterfly effect starts to make a little more sense.

During the time when Michael Fagenbaum was thinking about chaos, another scientist was thinking about a similarly chaotic principle. Although not a traditional physicist, Edward Lorenz was a meteorologist by profession but a mathematician at heart. He remained a meteorologist because he loved studying the weather and was utterly intrigued by its chaotic changeability. He understood that there was a randomness to the weather and it was always changing and never repeatable.

Lorenz built many weather models which showed that, when put in a natural state, weather patterns were never exactly repeated. They could be similar, but there were always some notable differences. He realized that, if a weather system was perfectly replicated in the world from one moment in time to another, the element of chaos would become null and void. If a specific weather system were to be perfectly repeated, then the following events in the weather system would eventually occur exactly the same way they have in the past. Then, the world's weather system would start to run on a continuous loop, even if that loop was millions and millions of years long.

Lorenz realized and understood that, by implementing all of the Newtonian laws into models revolving around nature, all of the randomness in the world and the universe could potentially remain "stable" for eternity. This brought about for Lorenz many thoughts about a "Higher Power" that designed the constructs of the universe.

Because of the randomness of the weather systems, Lorenz realized that there are so many different types of external factors that can play a role in the chaotic effects of weather systems. This also meant that long-term weather forecasting was practically impossible— not to mention largely inaccurate.

Short-term weather forecasts can still be relatively accurate, but anything beyond a week becomes complete guess work because of probable changeability, and Edward Lorenz knew this. He understood that there were far too many factors that could alter weather and certain courses over time. While performing his experiments, he noted that accurate input into a system will give an approximately accurate output value. Small changes at the beginning of either an experiment, or an actual weather system, will lead to massive changes along the way.

As a weather forecaster and meteorologist, he knew that weather patterns could potentially repeat themselves, but would never be exactly the same. Weather is a pattern with disturbances—it is an orderly form of disorder.

Lorenz knew that slight deviations in data would result in significant changes along the way. But, he didn't know how severe these changes could be until he designed a weather model to randomly assign numerical data during a certain time period. One day, he plugged in the numerical data and let it run for the evening before he left for home and came back the next day to assess the results. For the sake of repeatable accuracy, he decided to run the test again using exactly the same numerical data.

When he returned the following day after the test was completed for the second time, he noticed that the results were completely different from the first results. He figured that the machine must have malfunctioned during the night because these large types of fluctuations were not conceivably possible. After assessing the machine, he realized that it had not malfunctioned. But he was still struggling to fathom how such great deviations could have occurred in the results.

Upon closer inspection, he suddenly realized the truth of the experiment: In the first test-run, he included the numerical data according to six decimal places (specifically 0.506127) and let the test run on values with six decimal places. But because he was running late the next day, he only included the first three decimals of the numerical data. We have all learned that rounding off while doing math at school is alright... Little do we realize that it may have serious consequences for accuracy!

He closely analyzed the data and thought that, even though he rounded the data off to three decimal places (specifically 0.506), it wouldn't have such a large bearing on the accuracy of the experiment. So he started studying it further. He graphed the data of the two experiments, and noticed that both graphs started off exactly the same. But, because he placed the machine on a setting to randomly configure data after the beginning, the multiples of the decimal places eventually played a huge factor in the ultimate outcomes of the experiments.

Once the experiments were finished, Lorenz noticed that the end results were entirely different from one another, even though they both started from the same point. The deviations occurred because of small differences, but those small differences exponentially grew into something much larger and much more different. This was the birth of the theory behind the Butterfly Effect.

One of the most famous quotes that Edward Lorenz ever said was, "mathematicians seem to have no difficulty in creating new concepts faster than the old ones become well understood". This is far too true, and one of the downfalls of both modern and archaic science and mathematics. Before fully understanding a concept, there seems to be a race among many scientists to develop new ideas and learn new concepts faster than their peers.

Edward Lorenz was a mathematician by heart. So, when he noticed that there was a chaotic stability in the weather systems he was studying, he knew he needed to understand it. By choosing to first understand how chaos worked, he started understanding the deeper meanings of quantum physics. He first attempted to understand the old before he attempted to understand the new.

After he saw the results from these experiments, he instantly knew that long-range weather forecasting would be forever inaccurate, no matter how advanced the technology was. Because no matter how advanced technology can be, there are simply too many variables to consider to accurately predict what would influence the weather in the long run.

Something as small as a butterfly flapping its wings can (in theory) set off a set of events that would result in something far greater along the way—even something as great and powerful as a hurricane.

Chapter 3: Why Learn About Quantum Physics?

You should have realized by this point, that quantum physics is not something that you would be interested in if you weren't curious enough, by nature, to want to explore the deeper meaning of the universe. You want to know about the deeper functioning of the world around you.

The main reason to want to learn more about quantum physics is because it has a direct bearing on your life. Far too many people complain about having to take math as a subject at school. They say things like, "I will never use algebra in my life! So why do I have to be forced to take it as a subject?" One reason is that it allows your brain to think in certain ways and have the ability to solve more complex problems.

Quantum physics, however, does have a more direct impact on a person's life. Understanding how these different aspects affect us, can allow us to begin to understand the greater picture of the world that we live in.

We already know that both chaos and the butterfly effect directly influence our lives because of the different types of behaviors that people exhibit on a daily basis, for example, or the types of weather patterns that can occur without warning. Once a person starts to understand that there are billions of different variables that can occur in the orderly chaos of our lives, then they will understand that some things can happen without warning.

If you were the type of person that enjoyed *The Matrix* trilogy, then you will understand that there are very interesting types of theories out there in the world today. That said, the chances of us living in a largely detailed and coded program like *The Matrix* is rather unlikely, but there are still some phenomena that occur today which need to be taken into consideration.

In *The Matrix* movies, there were instances where "glitches" would occur and people would experience moments of déjà vu or other weird encounters. These moments of strange occurrences in the coding of the matrix itself made for stranger life experiences. There are countless Reddit and other social media posts that speak about the "glitch in the matrix" scenarios that occur to people in real life—some strange events that weren't deemed possible and are otherwise, unexplainable.

Since quantum physics allows for rather ambiguous and strange theories, it does make sense that there is a probability of other dimensions existing. Since it has been observed and theorized that particles can exist in other states at the same time, it is also plausible to consider that the very same particles can also exist in different dimensions at the same time, too. This type of discussion will be continued when we detail the theories and experiments of Erwin Schrödinger, and of course, his cat.

Though unlikely, these theories are important to note because there is a probability of other dimensions that affect us. There are other possible explanations like hallucinations due to exhaustion or panic, or there is something else that could have occurred that altered time and space in our dimension. These examples are included to help open your mind to the idea of concurrent dimensions, and that time warping is potentially possible.

The Disappearing Car

John Greene retired from the Army Rangers when he was around 32 years old. He had been injured during one of his tours to Afghanistan and was honorably discharged. Soon after, he received treatment and was on the road to recovery. He had trained and served as a medic during his time in the Rangers. Once he had recovered from his injuries, he still wanted to work, and thus committed enough time and effort to completing his studies as a paramedic in Colorado.

Because of his previous training, he obviously excelled and performed well with the crew at the local fire department. One evening, when Greene was heading home from work, he noticed that a car was pulled up on the shoulder of the road on the La Veta pass. Even though he was driving his personal vehicle, he pulled up behind the car. He always had a treatment bag and radio in his car in case he ever needed to treat victims in a crisis or to put in a call for help.

He was about 30 yards away from the vehicle and noticed that there was neither movement nor light coming from the car. Greene slowly exited his vehicle and walked towards the car with his flashlight in hand. He could see two people in the front of the vehicle but neither one of them were moving. Instead of walking closer to them (because it can be dangerous to do so), he merely called out from a distance. There was still no movement in the vehicle.

Greene then decided that it would be more prudent to call for backup and then approach the vehicle, just in case something happened to him. He moved back towards his vehicle and contacted the police station. Because of his time as a paramedic, he had become close friends with many of the officers, and they were more than happy to send assistance to help Greene.

Greene informed the officers that the car looked very suspicious and asked them to approach without their lights and sirens on. He climbed into his vehicle to get the GPS coordinates from his GPS device, but when he looked up, the car in front of him had vanished. He knew it didn't just drive away, because the section of the pass was completely straight and downhill. He could see along the road for miles. There was no other car on the road except for his own.

After several seconds of befuddlement, he cancelled the request for any backup to be dispatched. He walked over to where the car was a few seconds prior and there was no evidence that it was ever there. There were no tire tracks in the dirt on the shoulder of the road; there were no skid marks; there was nothing that could indicate that there had been a car parked there.

To this day, Greene still can't explain what happened out there on the La Veta pass. So, he decided to share his experience on Reddit, and many people have claimed to experience similar phenomena.

The Zanetti Train

This is one of the most legendary "ghost train" stories to ever be told. It occurred over a century ago, and people can still not explain what happened or why the strange occurrences appeared afterward. But, if this story is true, it may hint toward the existence of not only time warping, but different dimensions having access to different moments in our history.

In the summer of 1911, a new locomotive and railway contract was signed by the Italian government. To celebrate, one of the engines, named Zanetti, would leave Rome and go on a scenic trip in the Italian countryside. All three passenger cars attached to the locomotive were railway staff and their families. There were approximately 150 people on board.

The journey started without incident and everyone on board was in a very happy and excited mood. But, these things started to change when the train approached the tunnel in Lombard. People on the train started to complain about a humming noise that seemed to grow in intensity every few seconds. This was then accompanied by an unexplainable haze in the passenger cars. As things worsened, many of the passengers began panicking.

Passengers threw themselves towards the exit doors, but this only worsened the situation because the doors didn't open while the train was moving. However, two passengers managed to squeeze through openings in the train and dive off before the train entered the tunnel.

The two passengers, although hurt from the fall onto the hard ground, were well enough to go and seek help. They watched the train thunder into the entrance of the tunnel. It took several hours for them to locate help, but when Italian authorities did arrive, they contacted all of the stations on the line the train would have passed, but none of them saw the locomotive. They presumed the worst, that the train had crashed in the tunnel in Lombard.

Search parties went in at the entrance and at the exit to approach the wreckage from both ends to look for survivors and start stabilizing the crashed train. They were obviously very surprised when they met each other in the middle of the tunnel with no sign of the train ever passing through there.

The story only gets weirder from here. Even though there was an extensive search party formed to look for the train, it was never located. The Italian government, in a panic, decided to quash any reports of this disappearance as quickly as possible. But this didn't hinder the loved ones of the missing to continue the search. But for years, the train was never located.

The train would not be seen for the next 80 years, that is, until a strange, steam locomotive pulling three passenger cars was seen in the Ukraine in 1991. A railway worker in the small town of Poltava saw this strange train rumble past him as he worked one evening. He noted that the curtains were drawn over the windows, and he couldn't see anyone on board. The train, again, disappeared without a trace.

Other reports started surfacing when people started digging into this mystery. A recorded incident in

Mexico, in 1926, seemed very interesting, when 107 Italians appeared out of nowhere and appeared very disoriented and confused. They kept claiming that they had come from a train, but after the Mexican authorities investigated their claims, they could not find any trace of this train, nor a passenger manifest for these Italian passengers.

Then people started to discover that, as far back as the 16th century, Catholic monks in Italy reported a steaming beast with three parts being pulled behind it. Their ancient documents refer to the train as a beast that billowed smoke and rumbled along the countryside. It would then disappear without a trace, and never be seen again.

This report may be nothing more than a ghost story, but if one considers this from a quantum physics perspective, then it might be worth looking into a little more deeply. Alternate dimensions are plausible, and although there were only two that were mentioned in this chapter, there are thousands of other experiences just like these.

If particles can exist in other states and other dimensions at the same time, then it is a probable fact that, according to quantum physics, there might be a possibility for dimensions to coincide and alter one another.

Chapter 4: String Theory

If you thought that quantum physics and chaos were controversial topics, then string theory might just take the cake with this one. String theory, one of the most untested and unproven areas in physics, may just bring about more questions than it ever will answer. But it is a beautifully compelling theory that all physicists learn about. Fortunately, you won't have to be Sheldon Cooper, from *The Big Bang Theory,* to understand the basics of this theory.

This theory could potentially change the way that theoretical physics solves problems in the quantum realm. It can reduce multidimensional problems down to a singular dimension—making it much more likely to be solved in the long run.

At the core of this brilliant theory lies a thought that has been influencing physics for centuries: At some basic level or foundation, all beings, forces, interactions, and manifestations of reality are attached together and tied into one another in the same fundamental framework.

In physics, there are four fundamental forces that can influence a body in motion or at rest. These four forces are: Strong forces, weak forces, electromagnetic forces, and gravitational forces. In most areas of physics, these forces are normally separated from one another, with only one or two overlaps. But, string theory is the theory that unifies and encompasses all of them at once.

Many physicists believe that string theory might be the strongest contender for adequately describing the quantum theory of gravitation (the quantum theory that amalgamates at the highest forms of energy levels). It is true that string theory remains untested, but it does have a strong theoretical backing which sounds very plausible.

Ed Witten is one of the forerunners in string theory and has made very compelling arguments for his cause. In his most recent article *What Every Physicist Should Know About String Theory*, he described how many similarities can be noticed in all of the laws of nature. There seems to be all of these unrelated phenomena, but they all correspond and overlap somehow.

He noticed that two large planetary or orbiting bodies work with gravity and with one another according to Newtonian laws, exactly the same way that electrically-charged atoms either attract or repel one another. Also, the way pendulums can oscillate and swing, works with exactly the same principle as how a planetary body can orbit a larger star or a black hole.

Waves share several similarities, regardless of what type of waves they are. Water waves, light waves and gravitational waves all work in similar fashions, even though they are produced from entirely different sources. Because of these similarities, Witten realized that if a person wants to dive into the quantum theory of gravity, they will first need to grasp the quantum theory of a single particle. The reason for this is simple: Because it's the same fundamental theory and will work in the same way.

The Quantum Field Theory

Diving into quantum field theory is a very complex subject, but one can still grasp the basics of this theory with beginner-level knowledge. Quantum field theory requires a scientist to carefully study the behavior of a singular particle and solve the sum of its mathematical data over its histories. This is rather difficult, indeed—it is practically impossible. Calculating where a particle was and how it behaved over a certain time period may fluctuate because of external stressors. There will always be a foundational quantum uncertainty in nature, and all of this uncertainty can make it practically impossible to track a particle's history.

The solution would be for a physicist to solve for the sum of all of the possible ways that a particle reached a certain state. This would take some time since there are several calculations in probability that need to be performed to solve for the answer that is most likely the correct one.

Quantum field theory does not take Einstein's theory of relativity into account (which will be discussed in the next chapter) because the theory of relativity involves studying space-time geometry instead of the behavior of particles.

Quantum studies involve problems and solutions in all three spatial dimensions known to man. There are discussions and experiments regarding fourth and higher dimensions, but these are very advanced. For now, it is important to understand that we live in a three-spatial-dimensional world.

It is important to know that solving problems in all of three spatial dimensions can be incredibly difficult, even for the most experienced physicists. So, it is much simpler to take things down to a single dimension to solve these problems. This is why employing string theory to solve these multidimensional problems works so well. One of the few surfaces that continues on to include all of the problems in the different dimensions is a string. This string can have two open ends, or it can be considered a closed loop. Using the idea of a string to solve these problems has allowed for much greater probability of solving "unsolvable" problems.

Another additional benefit to employing string theory to solve these problems, is that spatial curvature in all three dimensions becomes impossible to solve, but once the dimensions are reduced to a singular dimension, then the spatial curvature problems become obsolete and easily solvable.

Reducing the dimensions to a singular dimension allows for "unsolvable" problems to be quantifiable. And, even though a particle may gain extra freedom while having access to all three dimensions, this extra freedom doesn't influence the outcomes of particle behavior in studied systems. There can be some inaccuracy if physicists aren't careful during this process, but if they define the particle's velocity as a vector quantity (which can work in both the first and second spatial dimensions), the results will remain accurate within the main dimension that is being studied.

If the problems remain within one dimension, then the problem of quantum gravity can be as easily solved as that of a single particle. This is not easy because there are many variables to consider, but it is much easier than trying to solve problems in the entire system.

String theory does offer a potential path to solving problems regarding quantum gravity. It can incorporate all four fundamental forces in a singular dimension to solve problems pertaining to the universe.

Chapter 5: Einstein's Theory of Relativity

Einstein was one of the greatest scientists in his field, and he managed to develop two theories of relativity that are greatly important to understanding quantum physics. His theories of general and special relativity are two of his most famous theories. Both general and special relativity revolve around gravity, how it influences physical phenomena and how it relates to other aspects and forces in nature.

General Relativity

Albert Einstein published an article in 1905 about how the laws of physics are all the same for observers that are not accelerating. He then further added that the speed of light (when placed in a vacuum) was completely independent of the motion of external observers. This is where the idea that a train approaching the speed of light will warp time, and time will slow down within the train but will remain the same for external observers. His article on general relativity introduced new concepts and ideas that would revolutionize different concepts of both space and time.

Einstein then spent the next decade trying to incorporate acceleration into his theory and published *General Relativity* in 1915. In it, he discussed how massive objects in space caused a type of distortion in space and time known as "gravity".

In the movie *Interstellar*, astronauts managed to get to another galaxy through a wormhole and looked for another planet to sustain human life. When they approached one of the planets (that was orbiting a large black hole), they realized that time on that planet would be much different compared to life on Earth. They concluded that a single hour on the planet would be equivalent to seven years on Earth. This concept flummoxed audiences, but according to Einstein's article on general relativity, it makes perfect sense that massive objects in space would be able to warp time and space if they had a large enough gravitational pull.

When two objects orbit one another, they will exert a force on one another known as "gravity". The three laws of gravity were designed by Sir Isaac Newton when he managed to quantify the gravitational pull between two objects. The ultimate strength of the gravitational pull is wholly dependent on the size of the bodies and how close they are to one another. This gravitational pull does not only include planetary bodies or larger orbiting masses; it also includes bodies that are on Earth at this moment. The center of the earth pulls everything towards it, and since you are on the Earth's surface, you will be planted firmly on its surface as you are being drawn toward its core.

Obviously, the gravitational pull of the earth isn't strong enough to pull you into the ground but is strong enough to keep you planted. It is also important to remember that you are also releasing a gravitational pull from the center of your body.

That seems strange, but a person's body does release a certain amount of gravitational pull. That said, a person's gravitational tug would be so minute in comparison, that it wouldn't make any difference to

Earth (even though there are more than seven billion people living in the world). Thankfully, we are firmly set onto the Earth's surface because of its gravity, even though this force can work innately and can act over a distance in most cases.

Einstein realized that the laws of physics are the same for any external observers that aren't accelerating. This means that the speed of light in a vacuum will remain the same regardless of how fast an external observer is traveling. For this very reason, he concluded that both time and space are carefully interwoven with each other in a single continuum known simply as the space-time continuum.

In quantum physics, one event can occur at a specific time for a certain external observer, but the same event could be witnessed at a completely different time by another external observer. This became apparent to Einstein when he realized that massive objects in space can distort and warp the space-time continuum. This is obviously not something that can be experimented on, but there have been several observations that have been made in space related to the warping phenomena.

By incorporating acceleration into his theory, Einstein managed to understand that time, in space and on other planets, may be completely different from time on Earth. It is important to keep in mind that, because of the differences in gravity, the space-time continuum can be influenced.

Special Relativity

Before Einstein realized that he should incorporate acceleration into his idea, he carefully studied space and light to understand how it would affect time if the gravitational pull of a specific object was altered in any way. During his observations of this theory, he concluded that if any object approached the speed of light, it takes on an "infinite" trait and is unable to travel any faster. Light is one of the most commonly associated and discussed factors in physics, and even in science fiction (especially when beings are trying to cover exceptionally far distances over galaxies).

Einstein's thoughts on special relativity eventually paved the way for him to make more substantiated claims regarding gravity, acceleration and the warping of space-time. He built on the work of previous astronomers, who built their theories on the three laws of motion according to Newton. After years of observation, trial, and plenty of errors, he eventually manufactured his famous equation for special relativity: **$E=MC^2$**.

This is one of the most commonly seen equations in the world today, but it is not an equation that many people know how to use. The equation is simple enough, however: $E=MC^2$ can be broken down into basic elements—that means that if you want to solve for energy (or mass because they are interchangeable in this formula) in the space-time continuum, then you need to multiply the mass of the object with the speed of light squared.

This formula can be used in any area that is influenced by gravity and by the speed of light. This formula also shows why the atomic bombs that were dropped on the cities of Nagasaki and Hiroshima were so enormously powerful. The atomic bombs that were dropped stuck closely to the laws of relativity that governed them. Since energy and mass are interchangeable in this formula, it meant that the bombs literally had four and a half tons of energy to release because of their splitting atoms inside.

Once this equation was adapted into the space-time continuum, Einstein saw that an object's mass would increase in space when it was exposed to the gravitational pull of other objects. This is why light, inside of a vacuum, becomes infinite and the fastest speed known to man. As any object moves through space, it's mass increases. So, if an object approaches the speed of light, its mass becomes so large that it approaches an infinite weight. (Which is why an object can't move faster than the speed of light because its mass has reached its ultimate limit).

On an additional note, the only reason why light can move as fast as it does, is because it actually does not have any weight. The photons, and the quantum particles that bring light into being, are considered weightless. This is why it is the fastest speed known to man. It has no weight to become infinitely heavy, so there is nothing holding it back.

Relativity has always been a strange and foreign concept to most of the world's population. Most people have heard of it, but they don't really know how to explain it. It's like asking someone to explain the concept of "time." It sounds easy enough, until you try it, then you realize that it's a lot more difficult than you realized.

Relativity is exactly one of those areas that are difficult to explain. Einstein described relativity as time that moves relative to the person being exposed to that time. In other words, you'll age less when moving than if you are at rest. This is why astronauts age more slowly than humans on Earth. They are not exposed to the Earth's gravity and are constantly moving at fast speeds. This phenomenon is known as "time dilation".

Relativity states that the faster you move, the slower you will age. If, for example, an astronaut leaves the planet when he is 25 years old, and travels at the speed of light for five years before returning to the planet, he would be 30 years old, whereas everyone else in his class would be about 70 years old.

Time dilation occurs at extreme speeds and gravitational manipulation, but that is what the theory of relativity is actually all about.

Chapter 6: The Bohr-Einstein Debate

Quantum physics is one of the most awe-inspiring fields of science. The only problem is that with these mind-boggling ideas, many people struggle to agree with one another. In other scientific fields, debates almost always end with an experiment or conclusive research, but the quantum side of things doesn't always allow for these types of "conclusive" conclusions.

Almost a century ago, two of the main "brains" behind quantum theory started butting heads because they couldn't agree on who was right. Niels Bohr and Albert Einstein went head-to-head to try and prove that they were right in several heated debates. There were, of course, other scientific members at this debate, but Bohr and Einstein definitely dominated the floor.

The Solvay Conference was held in 1927, in Brussels, where twenty-nine of the world's most influential scientists gathered to argue certain points about the newly developed quantum theories. But, it did not take long for the arguments to start because, right off the bat, Bohr argued that electrons, particles and other entities could only be accurately probabilistically measured if they weren't observed.

Einstein quickly got up and responded that all of those particles have realities that are independent of one another and coined his very famous phrase: "God does not play dice with the universe." The argument continued for some time afterward.

Unfortunately, in a field as complex as quantum physics, all the phenomena that occur are not normally fully understood because of the different types of variables that can externally influence them.

Quantum physics will always baffle people—even the best of minds.

Bohr continued his argument about wave-particle duality. He said that even though it can act as a particle, it is unlikely to do so because particles (like electrons) interfere with any light emissions or waves. He said that a system could act as either a particle or a wave, but it would be impossible to predict which one it would ultimately act like.

Heisenberg decided to get in the action at that point (since Einstein and Bohr were still at each other's throats), and he started discussing his uncertainty principle. He pointed out to Bohr that basic physics will influence a particle's momentum and position in any system.

This debate carried on for a much longer time afterward, and there are many questions that are still unanswered. Quantum physics simply doesn't allow for fixed testing in many cases and relies heavily upon conjecture and theories. There are still many views about this debate, even though it occurred in the 1920s. Most of the textbooks on quantum physics published after this debate sided more with Bohr's viewpoint and argument.

The Solvay Conference became one of the most famous debates of all time. And, even though the scientists that were arguing their viewpoints were all on the same standing, they ultimately believed in different paradigms. Bohr argued from an instrumentalist view while Einstein argued from a realist one. They brought up different variables that could have influenced the outcome in any quantum system.

There are, of course, other scientists that fully believed in Einstein's viewpoint and agreed with his argument. The Bohr-Einstein debate was one that caused much conflict and discord among the scientific communities, but it was still important.

This debate had two sides, and there were different viewpoints on both of the arguments. But, because much of the quantum world still remains theoretical, many of the arguments made are still considered just that—theories. There have been some advancements and experiments that have been conducted over the years to prove or disprove certain arguments, but there are still many areas of quantum that are still unquantifiable.

Both aspects of the debate had their merits, and were both important. Bohr and Einstein both managed to change people's ways in thinking and considered quantum theory as a viable scientific field.

Chapter 7: Quantum Entanglement

This chapter is included even though there are many scientists that believed that Albert Einstein and John Bell were wrong in their theory of quantum entanglement. That said, it is not definitively incorrect and may still be very valuable to many studies in quantum physics. Bell was the first to introduce this theory because he believed there was a possibility to warp the space-time continuum and allow for information to be communicated faster than the speed of light.

Earlier, it was stated that there is no conceivable way that anything can ever travel faster than light, but quantum entanglement can actually change that possibility if it turns out to be true. In this theory, Bell and Einstein claimed that two particles can be intimately linked from a very far distance, even if that distance is billions of light years away. And if one particle changes, the other will change in the same way. For this theory to be correct, or even possible for that matter, there would have to be areas between the particles that are warped to bring them closer together.

Very Spooky Action From a Distance Away

Einstein did struggle with this notion at first because he didn't believe that anything (even information) would be able to travel faster than the speed of light. In the last 50 years, several physicists have tried to find a way to transmit information faster than the speed of light. But, this has not proven to be possible yet because actually building a device that was capable of even attempting this would not only be near impossible—it would be very expensive.

However, in 2015, three different groups of researchers and physicists hypothetically tested the entanglement theory, and the results were not completely negative. They all concluded that there is a little more basic support for this theory. In other words, there might be a way to transmit information faster than at the speed of light through looking at entanglement.

One of the primarily successful study groups was led by a physicist named Krister Shalm. Even though the study was expensive and rather unlikely to succeed, they decided to push through and attempt it anyway.

The team super cooled metal strips to the cryogenic temperature of absolute zero, -460 degrees Fahrenheit—which is the coldest that any temperature can be forced to. They can super conduct particles at that temperature because they do not have any electrical resistance at absolute zero.

The scientists then fired a single light photon at the super cooled metal strip (which changed the strip back to its normal conductive state for a split second), and the team then recorded the results. This allowed them to observe how photons would affect each other if they were part of an entangled pair.

The photons that were fired started responding like each other, even though they were fired at different times onto the super cooled metal strip. It is still difficult to know how far this sensitivity runs for. Considering the universe is infinite (probably), then it is still difficult to know if one photon on one side of the universe can result in entangled changes in another photon on the other side of the universe.

Even though the distance of billions of light years may be a problem, the scientists did notice that photons can, indeed, act together as part of an entangled pair. Shalm fully believed that Bell and Einstein were right in their theory and development of photon entanglement. This research (even though expensive) has given NASA enough data to employ into real-life testing and missions. They incorporated the data into their photon detectors and communication lasers.

After all of these years, this seemingly impossible theory may well be right.

Chapter 8: Schrödinger's Cat

Erwin Schrödinger thought of a paradoxical situation to get his point across regarding paradoxes and quantum superpositions. It is not believed he performed this on a real cat; it is considered a thought experiment.

Schrödinger's cat is an experiment that is based on a paradoxical thought or situation in which an organism can be in two states at the same time. In 1935, he decided to argue his point regarding a state known as quantum superposition. This is linked to a randomly activated subatomic event that simultaneously does and does not occur.

Schrödinger realized that there are two hypothetical scenarios when a living being can be both alive and dead simultaneously. It is a paradoxical situation of being both alive and dead at the same time.

His Thought Experiment

Schrödinger penned this hypothetical situation where one can place a cat in a controlled box. This box would be controlled and kept entirely separate from external influences. A Geiger counter would be placed inside the box to measure any possible forms of radiation that could be released from a controlled container inside the box. A small amount of a radioactive substance would be placed inside the box with the cat.

This radioactive amount needs to be large enough that it could potentially decay and have the equal potential not to. If the radioactive substance does decay and release a radioactive particle, a hammer would be released and break a bottle of poison in the cat's box, killing it instantly.

If one has left the entire system alone for an hour with the radioactive substance's potential to decay, it can be argued that the cat can be both alive and dead at the same time, simply because one does not know about the state of the poison or the cat. This allows for a paradoxical state because the cat can be considered to be alive and dead within the same realm of reason.

But, this thought experiment does bring about a certain number of questions, with the main one being: When does a quantum system move from a paradoxical state or a state of superposition and concede to the actual results of reality? In other words, when does a quantum state stop being an amalgamation of states, each of which could be considered classically real states? And then move into an area of a single state or description?

If the cat survives this dangerous experiment, then it will only remember being alive (because it didn't actually die). But this poses a problem because, according to quantum physics, both dimensions need to run concurrently for the sake of accuracy. Quantum physics may run in paradoxical states, but alternate dimensions and realities are an important factor in this field.

This thought experiment did open the door to further studies of quantum superpositions. The problem that Schrödinger encountered was that he could not ascertain whether the living cat was merely an observer of this experiment—or did it become part of the classical state and another external observer needed to be included?

Superpositions

Since Schrödinger brought up the possibilities of superpositions, many scientists have explored the possibilities of these paradoxical states. The only problem that they all have encountered is that they don't know how long they're supposed to last and collapse (if they ever do collapse).

The first interpretation that was used for superpositions was the Copenhagen interpretation. This is the most common form of interpretation because the system stops being a superposition as soon as the results are observed. This would allow for one state to be accepted. In the case of the experiment with the cat, though the cat would have been both alive and dead in the box, its superposition would have ended as soon as the box was opened. This meant that thought experiments needed to be ultimately measured or observed according to the Copenhagen interpretation.

This became a common trend among many physicists but there were a few that disagreed with this notion. Niels Bohr was one of them. He didn't agree with the thought of this type of superposition and thought that the cat would have been dead or alive long before you observed it. He implied that there was only one reality for the cat in the experiment, and it had nothing to do with a conscious external observer.

This theory of superpositions raises the question of whether there are potentially other dimensions that are being lived in at this very moment somewhere else in the universe? If superpositions are enhanced and increased in size to include other dimensions running concurrently with the one that we are presently witnessing, then does that change the concept of the universe into a multiverse? And, does this concept allow for other overlapping of dimensions?

These questions are difficult to answer for the very reason that other aspects in the Bohr-Einstein debate were difficult to side with, and there aren't quantifiable results (yet) to justify these claims. It does make sense to "theoretically" believe that there are other dimensions and parallel universes running concurrently with ours, with the overlapping scenarios that could possibly occur (like in the "glitch in the matrix" section).

There are reasons to believe that superpositions are real and that we are being exposed to many of them in our current dimension. There are also theories that we are only observers of our current dimension and might be living in other dimensions because we ourselves have entered states of superpositions. Fortunately, those types of theories are normally reserved for extremists in quantum beliefs. It is still an interesting thought, however.

It is important to consider that there are debates about superpositions and their interpretations. And there were several other (less common) interpretations that were designed along the way. Even though there are several scientists that don't agree with the idea of superpositions, it does allow for many future debates regarding larger organisms in multiple states at once.

Chapter 9: The Double-Slit Experiment

Some people have regarded the double-slit experiment as one of the most beautiful experiments ever performed by physicists. This experiment showed that both matter and light are capable of displaying characteristics of both waves and particles. Also, it was a pretty novel experiment because it was performed long before there was a traditional form of physics.

This experiment was first performed in 1801 by Thomas Young. He wanted to demonstrate that light could behave in different ways if put into a certain system. (Quantum physics didn't exist back then, but he knew that there was a much bigger field in science that was on the precipice of being born). This experiment predated all other quantum experiments, but still remains as one of the most influential.

Modern physics was developed over a century later, but because of his research, they quickly knew that light could act as both particles and as waves. This experiment was the first of many to split a wave into two separate waves.

The Basic Version of this Experiment

For this experiment to work, there needs to be a strong and coherent light source or laser beam that illuminates a flat object that has two parallel slits cut into it. The light passing through the slits is then observed to identify what type of nature the light is exhibiting.

Since light was first understood to stick to its wavelike nature, then it was assumed that one would see black and white strips of light passing through the slits...

Which people did. The light responded exactly as the scientists thought it would and did show black and white bands. However, they noticed some behavior they did not expect. The light also exhibited the same behavior as the nature of particles. There seemed to be dimming areas on the sides of the slits that showed that the light also responded like particles.

This experiment was then increased to much larger proportions than merely photons and electrons. That said, increasing the size of this experiment makes it much more difficult for successful results.

The variations of the double-slit experiment have become very commonly seen thought experiments. It is known to clearly express problems and central puzzles of quantum mechanics. Richard Feynman said that this experiment borders on the impossible because of its phenomenon that responds in a classical way. It contains the mystery of quantum mechanics within the two slits in the experiment that the light passes through.

It may have been a very simple experiment to perform, but it was one that changed the outlook of many scientists in many scientific fields.

Chapter 10: Time Travel

Every lover of science fiction will tell you that they adamantly believe in the possibility of time travel. Many movies include wormholes in the mix to move from one dimension to another, or from one time period to another. After all, it is part of many different science fiction movies, but the real question is, is it actually possible in reality?

According to Einstein's principles in relativity, it might actually be possible. But, the probability of this may be rather unlikely. The reason for this is that our understanding of the dynamics of the universe still leaves much to be desired. For time travel to be possible, general relativity needs to be fully grasped. Although the theory of general relativity does predict how large objects like stars and planets interact, it cannot fundamentally explain the deeper functioning of the universe.

General relativity only focuses on the macro world and how gravity and speed can warp the space-time continuum. But, it doesn't take into account the smaller chaotic variations of quantum changes.

Because there are such large distances in the universe, there can literally be an infinite number of galaxies. Just as a sequence of numbers, there is no finite end to the universe (that we know of).

Before any thoughts of time travel can be made, it is first important to understand what would be needed to actually travel through time. Traveling through time would require a person to travel vast distances in a very short space of time while warping the fabric of the space-time continuum. If a person could leave an airport and fly around the world, they could land at the same place, but at a different time. For time travel to be beneficial, it needs to be performed in a dimension that is removed from the normal processes of Time that governs our world.

Wormholes

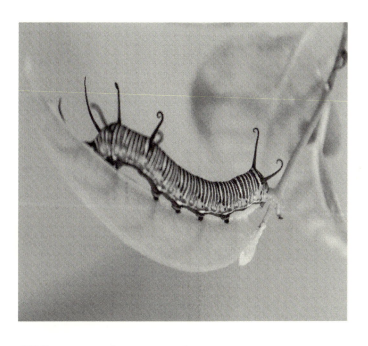

(Well, not exactly a worm…)

In quantum physics and theories of quantum gravity, the only way a being would be able to travel faster than the speed of light would be for a person to go through a wormhole. Wormholes are tears in the space-time continuum that allow for speeds potentially faster than the speed of light (according to quantum gravity). But it is difficult and confusing to understand that a wormhole is a tear in the space-time continuum.

To explain this, one should grab a piece of paper and draw dots on opposite ends of the page, then measure the distance between them. If the page is then folded in half, the dots would be right next to each other. This greatly lessens the distance that would have to be traveled between dots, and this is exactly how wormholes theoretically work.

Wormholes fold the distance of space and make it considerably less far to travel, and if one is traveling at astronomically fast speeds (hypothetically 95% of the speed of light) then enormous distances would be able to covered in a very small amount of time, and if there was an additional form of time dilation and gravitational pull, then a being might be able to time travel.

That said, this is only a hypothetical situation. A tear in the space-time continuum will require manipulation of fields in quantum physics that can only be dreamed of at this point. Further, wormholes aren't a probable thought yet for anyone that studies string theory. Though string theory makes it easier for problems in the quantum realm to be solved in a singular dimension, the only way for a wormhole to work is to manipulate dimensions three and higher. These are not aspects that are available or feasible at this point.

Traveling to the Future

This section has been included because, even though traveling through time needs a wormhole, there is another possibility. Hypothetically, if a person was planning on traveling to the future (and staying there because there wasn't a way back), then one wouldn't need a wormhole—only enormously fast speeds.

The previous sections in this chapter are not possible because of the unlikely possibility of wormholes being developed through external manipulation. But, there is a form of time travel that is available to humans now. Traveling to the future is possible but it might not be the type of future that people expect.

As discussed in previous sections, there are several aspects and factors that can warp time and space. Increased speed and gravitational pull will influence the time-factor and time dilation for the person experiencing it. Astronauts will age slower than people on Earth because the gravity they are exposed to is significantly less than the gravity on the face of the Earth. In addition to the decreased gravity, the speeds that are reached by being in space or orbiting a planet will result in much greater speeds than are possible within the atmosphere.

Some of the speeds that can be reached can exceed more than seven thousand miles per hour, which is nowhere near the speed of light. But it is fast enough to create some time dilation and slow down the aging process of the person traveling at those speeds.

Time travel and teleportation are unlikely events in this realm. There are simply far too many variables that need to be considered. And there is still far too much "raw data" that the universe hasn't offered up yet about understanding these types of events. Time travel and teleportation will always remain theoretically possible.

However, because we haven't been able to design anything with enough energy to warp the space-time continuum, we will be stuck in this time for the foreseeable future.

Quantum physics has opened up the possibility for them to occur, but there will have to be major leaps in understanding of this scientific field. At this moment, time travel is reserved for hypothetical cases.

Conclusion

Einstein had many great quotes in his life, but one of the greatest was when he said that "imagination is greater than knowledge." He said those words because he understood that the world and the universe are very complicated entities. The only way to try and understand the deeper meanings of them was to have a rather large and creative imagination.

He understood that quantum physics was going to require a crazier level of imagination than any knowledge that a book could ever give you. He figured that it doesn't make sense to limit your creativity and imagination because certain textbooks limit certain actions and reasoning. He knew that to fully embrace the idea of quantum, he needed to be able to expand his mind and imagine the concept of something much greater than himself, the world, the universe, or even the dimension that we live in.

For you to open your mind to embrace the concepts of quantum, you will need to fuel your imagination. Think of the unthinkable, and consider that things outside the realm of immediate logic can actually be possible.

Quantum physics is not normally reserved for beginners, but there are certain amounts of understanding that one can have for this type of advanced scientific field. It is true that you do need an extensive scientific background to understand deeper levels of quantum, but it is still possible for anyone to understand the basic functionality of this field.

There are so many different aspects about the world and the universe that people have taken for granted. People don't give enough credit or thought to the fact that we are living in an orderly state of chaos. There are an infinite number of variables that could occur in the world on a daily basis. The world currently has approximately seven billion people in it. If every person makes 100 decisions every day, then there are practically 700 billion different outcomes that can occur in the world on a daily basis. And that's just in the world. That doesn't take into account the trillions of possible outcomes that could occur just in our solar system, let alone the rest of the universe.

Those are just personal decisions. Some decisions that people make directly influence others and can alter another person's reality forever. It is sometimes an uncomfortable thought of how many different possibilities there are that can occur to a person on a daily basis. This orderly form of chaos that is our world is run by a deeper quantum fabric that we are still fighting to understand.

This book is filled with information that will change your view of the world around you, and the universe (or multiverse) that we all share. There are different dimensions that can possibly exist, and we may all be part of different states in different superpositions. There are phenomena that have occurred in the world (and are still occurring today) that can't be explained by anything else.

Fortunately, there is a field that includes all of the weird and wonderful aspects of the world. Even though quantum physics is based on the tiniest particles known to man, these particles are the fundamental make-up of everything that we know in this world and in the universe.

There will still be countless leaps and breakthroughs that will be made in this quantum field of science—it is inevitable. The world is progressing at a speed not previously imagined. The world has progressed and evolved more in the last ten years than it has in the last century.

Even though many aspects of quantum physics may still remain theoretical, there are some breakthroughs that support previous claims because of progression in technology and other research avenues.

This book was designed to spark your curiosity and imagination for the world we live in and the science that strives to describe it. Fortunately, there aren't tons and tons of mathematical formulas for you to wade through. But, this is only the beginning step on a much greater journey. The world is filled with information ripe for the picking. It is up to you now to continue to seek an understanding of what is known and to imagine the unknown.

Hopefully, this book will be a small spark in your life that will spread into something much larger and brighter. The possibilities are limitless!

Bluesource And Friends

This book is brought to you by Bluesource And Friends, a happy book publishing company.

Our motto is **"Happiness Within Pages"**

We promise to deliver amazing value to readers with our books.

We also appreciate honest book reviews from our readers.

Connect with us on our Facebook page www.facebook.com/bluesourceandfriends and stay tuned to our latest book promotions and free giveaways.

Citations

Bailey, J. (2020). *White Yarn.* Copyright Free Image from Pexels.
   https://www.pexels.com/photo/white-yarn-745761/

Bartus, D. (2020). *Light Inside Chest Box.* Copyright Free Image from Pexels.
   https://www.pexels.com/photo/analogue-art-box-chest-366791/

Benton, J. (n.d.) *Shallow Focus of Clear Hourglass.* Copyright Free Image from Pexels.
   https://www.pexels.com/photo/shallow-focus-of-clear-hourglass-1095601/

Bikos, K. (n.d.) *How Do Atomic Clocks Work?* https://www.timeanddate.com/time/how-do-
   atomic-clocks-
   work.html#:~:text=How%20Accurate%20Are%20Atomic%20Clocks,the%20
   world's%20most%20precise%20clocks.

Black, P.E. (2002). *Quantum System: An Overview.*
   https://www.sciencedirect.com/topics/engineering/quantum-system

Carol, B. (2020). *White and Gray Cat in Brown Woven Basket.* Copyright Free Image from
   Pexels. https://www.pexels.com/photo/white-and-gray-cat-in-brown-woven-
   basket-1543793/

Dibert, D. (2020). *Brown Mouse Beside Leaf.* Copyright Free Image from Pexels.
   https://www.pexels.com/photo/animal-apodemus-sylvaticus-blur-brown-
   730469/

Dil, A. (2020). *Person Standing on Road.* Copyright Free Image from Pexels.
   https://www.pexels.com/photo/person-standing-on-road-2902747/

Emspack, J. (2016, February 14). *Quantum Entanglement: Love on a Subatomic Scale.*
   https://www.space.com/31933-quantum-entanglement-action-at-a-
   distance.html

Espacio, M. (n.d.) *Blue Pink and White Andromeda Galaxy Way*. Copyright Free Image from
Pexels. https://www.pexels.com/photo/sky-space-milky-way-stars-110854/

Figueras, P. *Car Passing on Road Between Trees*. Copyright Free Image from Pexels.
https://www.pexels.com/photo/bosque-car-car-lights-cold-626155/

Fernandez, E. (2019, September 9). *Could Quantum Gravity Allow Us to Time Travel?*
https://www.forbes.com/sites/fernandezelizabeth/2019/09/09/could-
quantum-gravity-allow-us-to-time-travel/#2b597c6d2f2a

Gleick, J. (1987). *Chaos: The Amazing Science of The Unpredictable. Making a New Science*. Vintage
Books.

Howell, E. (2017, March 30). *Einstein's Theory of Special Relativity*.
https://www.space.com/36273-theory-special-relativity.html

iStock. (n.d.) *Black and Gray Microphone*. Copyright Free Image from Pexels.
https://www.pexels.com/photo/black-and-gray-microphone-164829/

iStock. (n.d.) *Black Train on Rail and Showing Smoke*. Copyright Free Image from Pexels.
https://www.pexels.com/photo/black-train-on-rail-and-showing-smoke-
72594/

iStock. (n.d.) *Blue Morpho Butterfly*. Copyright Free Image from Pexels.
https://www.pexels.com/photo/blue-brown-white-black-66877/

iStock. (n.d.) *Person Pouring on White Ceramic Teacup*. Copyright Free Image from Pexels.
https://www.pexels.com/photo/caffeine-coffee-cup-drink-374780/

Kostak, B. (2020). *Man in Black Hoodie During Snow Weather at Daytime*. Copyright Free Image
from Pexels. https://www.pexels.com/photo/snow-winter-glass-frozen-
68917/

Lazy Masquerade. (2016). *10 Freaky Glitch in the Matrix Stories from Reddit*.
https://www.youtube.com/watch?v=mANDhM5rRto

NewScientist (n.d.) *Quantum Physics: Our Best Basic Picture of How Particles Interact to Make the
World*. https://www.newscientist.com/term/quantum-

physics/#:~:text=What%20is%20quantum%20physics%3F,biology%20work%20as%20they%20do

Orzel, C. (2015, July 8). *Six Things Everyone Should Know About Quantum Physics*. https://www.forbes.com/sites/chadorzel/2015/07/08/six-things-everyone-should-know-about-quantum-physics/#2f47060d7d46

PhysicsWorld. (2002, September 1). *The Double-Slit Experiment*. https://physicsworld.com/a/the-double-slit-experiment/

Piacquadio, A. (n.d.) *Woman Holding Books*. Copyright Free Image from Pexels. https://www.pexels.com/photo/woman-holding-books-3768126/

Raymers, M. (n.d.) *Photo of Person's Hand with Blue Light*. Copyright Free Image from Pexels. https://www.pexels.com/photo/photo-of-person-s-hand-with-blue-light-3345270/

Redd, N.T. (2017, November 7). *Einstein's Theory of General Relativity*. https://www.space.com/17661-theory-general-relativity.html

Redd, N.T. (2018, March 7). *How Fast Does Light Travel? The Speed of Light*. https://www.space.com/15830-light-speed.html#:~:text=The%20speed%20of%20light%20in,a%20lot%3A%20about%20670%2C616%2C629%20mph.

Siegel, E. (2016, November 25). *What Every Layperson Should Know About String Theory*. https://www.forbes.com/sites/startswithabang/2016/11/25/what-every-layperson-should-know-about-string-theory/#70eb329a5a53

Sinya, A.K. (2016). *The Train - Ghost. Mysterious Disappearing of a Train in a Tunnel in Italy*. https://steemit.com/story/@aksinya/the-train-ghost-mysterious-disappearing-of-a-train-in-a-tunnel-in-italy

Skibba, R. (2018, 27 March). *Einstein, Bohr and the War Over Quantum Theory*. Books and Art. https://www.nature.com/articles/d41586-018-03793-2

Uanthoeng, A (n.d.) *CLose-Up Photo of Assorted Color of Push Pins on Map.* Copyright Free Image from Pexels. https://www.pexels.com/photo/close-up-photo-of-assorted-color-of-push-pins-on-map-1078850/

Made in United States
Orlando, FL
03 December 2021

11077724R00079